40 Cool Science Tricks

# the Surfing Scientist

## 40 Cool Science Tricks

Ruben Meerman

ABC Books

For my favourite brats –
Luke, Hannah, Elly, Courtney, Samuel,
Gabriella and Romy.

The ABC 'Wave' device is a trademark of the Australian Broadcasting Corporation and is used under licence by HarperCollins*Publishers* Australia.

This edition first published in 2007 by ABC Books for
the Australian Broadcasting Corporation.
Reprinted in 2009
by HarperCollins*Publishers* Australia Pty Limited
ABN 36 009 913 517
harpercollins.com.au

Copyright © text and photographs Ruben Meerman 2007

The right of Ruben Meerman to be identified as the author
of this work has been asserted by him in accordance
with the *Copyright Amendment (Moral Rights) Act 2000*.

This work is copyright. Apart from any use as permitted under the
*Copyright Act 1968*, no part may be reproduced, copied, scanned,
stored in a retrieval system, recorded, or transmitted, in any form
or by any means, without the prior written permission of the publisher.

**HarperCollins*Publishers***
25 Ryde Road, Pymble, Sydney, NSW 2073, Australia
31 View Road, Glenfield, Auckland 0627, New Zealand

National Library of Australia Cataloguing-in-Publication

Meerman, Ruben.
    The surfing scientist : 40 cool science tricks.
    For children.
    ISBN 9780733320804 (pbk.).
    1. Science – Experiments – Juvenile literature.
    2. Scientific recreations – Juvenile literature.
    I. Australian Broadcasting Corporation. II. Title.
500

Internal and cover design by Tou-Can Design
Additional design work by Michelle French at French Curve
Colour separations by PageSet, Victoria

The ABC is not responsible for the content found on non-ABC internet sites.

# Contents

| | | | | |
|---|---|---|---|---|
| Introduction | | 6 | 25. Flippin' Coin Trick | 58 |
| Nag Nag Nag! | | 8 | 26. Science on a Shoestring | 60 |
| 1. Straw Pipette | | 10 | 27. Anti-gravity Coaster | 62 |
| 2. Density Layers | | 12 | 28. Magic Hanky | 64 |
| 3. Toothpick Star | | 14 | 29. Polarised Sunnies Trick 1 | 66 |
| 4. Pepper Scatter | | 16 | 30. Polarised Sunnies Trick 2 | 68 |
| 5. Psychedelic Milk | | 18 | 31. Phantom Patterns | 70 |
| 6. Food Colour Vortex | | 20 | 32. Paperclip Linker | 72 |
| 7. Balloon on a Stick | | 22 | 33. Orange Life Jacket | 74 |
| 8. Fireproof Balloon | | 24 | 34. Mysterious Egg | 76 |
| 9. Leakproof Bag | | 26 | 35. Wasabi Diver | 78 |
| 10. Spear a Spud | | 28 | 36. Corrugated Paper | 81 |
| 11. Super Can | | 30 | 37. Coin Paper | 82 |
| 12. Uncanny Cancan | | 32 | 38. Water Elevator | 84 |
| 13. Straw Atomiser | | 34 | 39. April Fools' Banana | 86 |
| 14. Levitation Trick | | 36 | 40. Slap on a Cap | 88 |
| 15. Sticky Stream | | 38 | | |
| 16. Inseparable Books | | 40 | Ocean Animal Facts | 90 |
| 17. Gripping Rice | | 42 | Beach Facts | 91 |
| 18. Mini Icebergs | | 44 | World's Weirdest Waves | 92 |
| 19. Ice Cube Lava Lamp | | 46 | Safe Experimenting | 93 |
| 20. Ice on a String | | 48 | Basic Science Trick Equipment | 94 |
| 21. Flying Cups | | 50 | Cool Websites | 95 |
| 22. Flying Tea Bags | | 52 | | |
| 23. Marble Graviton | | 54 | | |
| 24. Balloon of Death | | 56 | | |

# Introduction

The question I get asked most frequently is where did I learn my science tricks? Well, scientists always give credit for other people's work and ideas, so this book *should* be packed with acknowledgments and that's because I didn't discover or invent any of the tricks you will find in it... not one! But good science tricks are a lot like good jokes – no one ever remembers who came up with them first.

Consequently, we know that Louis Pasteur's breakthrough gave us pasteurised milk, but we have no idea who discovered that you can spear a raw potato with a drinking straw (see page 28 if you want to learn how). We know Sir Isaac Newton wrote the theory of gravity, but can't name the person who pioneered the art of lifting a full rice jar with a pencil (see page 42 for instructions). And we've all heard of Albert Einstein but who was the bright spark behind the incredible toothpick star trick (see page 14)? I wish I knew!

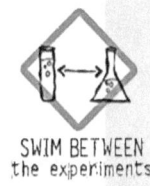

I can't name the anonymous people who first tried these tricks but I'm very grateful they did. Instead, I'd like to thank all scientists everywhere, for the invaluable work they do. They will give us new medicines, speed up the Internet and guide us towards a sustainable future. So if you're looking for someone you can really admire, forget those celebrity magazines and head to your nearest laboratory! In the meantime, happy experimenting... and surfing!

# Nag Nag Nag!

You don't need to stop bathing, dress badly or speak dolphin to be environmentally friendly! The small things we all do really can make a big difference. Here are some tips to remember while experimenting with the tricks in this book.

Pour leftover water from your experiments into a watering can.

Keep a bucket handy to avoid spills.

Then empty the bucket into a watering can to use on your plants.

Plants don't mind a little food colour, detergent or cooking oil in their water. I know because this plant told me so!

SCIENTISTS
patrol these waters

Recycle. And go nuts at people who don't! Nag and nag and nag until they can't bear it any more!

You can get away with nagging if it's for a good cause. Nag adults to use public transport or start a car pool.

Compost your food scraps, don't bin them. Then you can grow lots of worms and scare people with them! Ooooh, worms!

Don't just get angry about rubbish, pollution and global warming. Start nagging adults to do something about it.

If you *do* want to speak dolphin, try saying "micky! micky! micky!" in a loud squeaky voice...it sounds just like Flipper!

# 1 Straw Pipette

This nifty trick is also a handy skill to have up your sleeve when you can't find an eye-dropper. And let's face it, how often do you have an eye-dropper in your pocket?

Dip a straw into a glass of water. I've coloured the water so you can see it clearly.

Cover the end of the straw with your index finger and lift up. The water stays inside the straw!

Move the straw to where you want the water. Remove your finger and the water pours out.

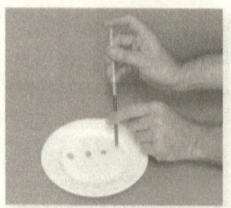

To release one drop at a time, start over but keep your index finger on the straw. Use your other hand to squash the straw and a single drop of water comes out... handy!

## What's going on?

Sandologist
ON DUTY

Air pressure combined with the surface tension of water make this trick possible. For water to leak out of a straw, an equal volume of air has to get in. Obviously, no air can get in through the top of the straw when your finger is blocking the way. But water molecules stick together so tightly that air can't squeeze in through the bottom either. This stickiness gives water its remarkable surface tension.

The other reason the water doesn't escape is air pressure. Air exerts pressure on everything in it. We don't usually notice this pressure because it is equal in every direction, including up. This rarely noticed but very hefty air pressure prevents the water pouring out.

But water molecules aren't infinitely sticky, so there's a limit to the thickness of straws you can use for this trick. Really fat straws won't work. Beyond their limit, water molecules break apart so air bubbles can sneak in. Then gravity wins and the water comes out.

### Pipetting Gizmos

▼ Pipette literally means 'little pipe' and that's exactly what a drinking straw is. Scientists use high-tech gizmos based on exactly the same principle to add precise amounts of chemicals into experiments. Turning a dial sets the volume of liquid to be transferred. I'm not sure what I'd use it for, but I want one!

# 2 Density Layers

Use your new-found pipetting skills to create cool layers of coloured salt solutions inside a straw.

Fill three plastic cups with water. Add red food colour to one, yellow plus one tablespoon of salt to another and blue plus two tablespoons of salt to the third.

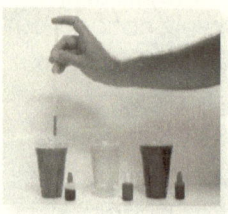

Dip a clear straw about 3 cm into the red cup. Seal the top of the straw with your index finger and lift. You should have about 3 cm of red water in the straw.

Keep your finger on the straw and dip it about 6 cm into the yellow cup. Release your index finger briefly, replace it and then lift the straw out of the water. Now you have a layer of yellow below the layer of red water.

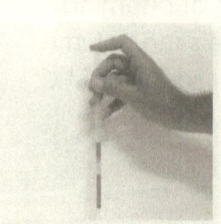

Dip the straw about 9cm into the blue cup and repeat the process. Lift the straw and you'll have three separate layers – red, yellow and blue. Pretty!

## What's going on?

serious experimentation underway

Adding salt changes the density of water. Density is how much a given volume of a substance weighs. One litre of pure water weighs exactly one kilogram. One litre of seawater weighs about 1.03 kilograms. The extra weight is due to dissolved salts and minerals.

The three solutions you made have different densities. The saltier solutions are more dense so the less salty solutions float on them. By dipping the straw in unsalted water first, then the saltier solutions, you get three distinct layers. It's amazing how little mixing there is between the layers. But try doing it in reverse order and you'll get lots of mixing and no distinct layers.

## Thermohaline Circulation

▼ The salinity (saltiness) of seawater varies quite a lot around the world. On average, there is about 35 grams of salt in every litre of seawater. Rain doesn't contain salt, so rainwater decreases salinity near the surface. Evaporation increases salinity, which can make the surface water slowly sink.

▼ To complicate matters, cold water is more dense than warmer water so it sinks too. These variations in salinity and temperature generate huge but very slow ocean currents around the world. Scientists call this the thermohaline circulation which is vitally important to the Earth's climate and sea life.

# 3 Toothpick Star

This trick will make you a star... literally! Use the Straw Pipette trick on page 10 to add the water if you don't have an eye-dropper. It works using matchsticks too.

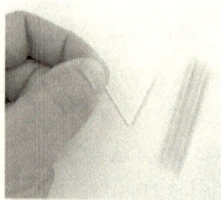

Carefully snap five toothpicks in half so the two halves are still joined and look like the letter V. Break them as neatly as possible.

Arrange the five toothpicks in a circle like this. The more neatly you arrange them, the better the result!

Gently squeeze a large drop of water into the centre so it touches the snapped part of each toothpick. I added food colour to the water but you don't need to.

Watch carefully. The toothpicks slowly make a beautiful star. Oooo-aah!

SWIM BETWEEN
the experiments

## What's going on?

You've probably figured out how this works, but just in case, here's a hint. What happens to wood when it gets wet? That's right, it swells! And that's what's going on here.

As the tiny bit of wood holding the two halves of a snapped toothpick together swells, the V shape opens up. Try it with just one toothpick and you'll see this clearly.

By arranging them in a neat circle, the tips of each toothpick collide with those of its neighbours. Once all the tips are touching, the star opens up like a flower. Cool!

The same phenomenon can cause old doors to swell and jam when the humidity rises. As soon as dry weather returns, the door opens freely again. It's a bit like the roof that only leaks when it's raining.

### Handy tip:

▼ If you learn to pipette water using a straw (see trick 1), you can do this trick in almost any restaurant. Who knows, you might impress the waiter enough to score a free dessert...yum!

Pointless FACT

Did you know that the first toothpick manufacturing machine was patented in 1872 by Silas Noble and J P Cooley of Massachusetts in the USA? What a coincidence...neither did I!

# 4 Pepper Scatter

This trick doesn't make any noise but show a friend and you'll hear the 'ooooh' sound of amazement in their voice. You'll need fresh water each time so have a bucket handy.

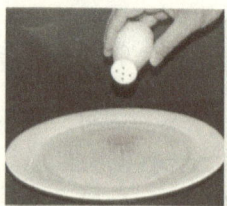

Sprinkle ground black pepper on a plate or bowl full of water.

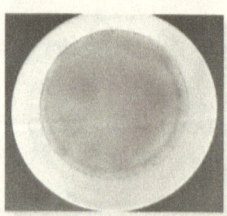

Cover the entire surface with pepper so it looks like the picture.

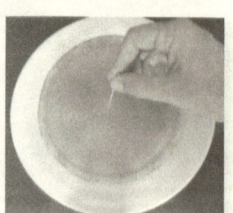

Touch the water with a clean toothpick. Nothing happens.

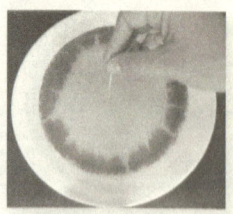

Now dip the toothpick in detergent and touch the water again. This time the pepper scurries away in all directions. Ooooh-ah!

SCIENTISTS patrol these waters

## What's going on?

Detergent molecules are amazing little things. They're way too small to see but scientists have figured out that they have two very distinct ends, called the head and tail. Their heads are strongly attracted to water but their tails strongly repel it. As a result, they prefer being on the surface with their heads in the water and their tails sticking out.

When you first add detergent to water, the molecules race across the surface with their heads down and tails up. The remaining detergent molecules form droplets under the surface by joining tails.

Detergent heads don't attach to water as tight as water attaches to itself. And this is why detergents reduce the surface tension of water. Imagine a long line of people all holding hands and pulling towards each other at the same time. The line is under tension. If a person in the middle lets go of both hands, everybody falls *away* from that person. The tension has been broken. In a similar way, water molecules on the surface pull *away* from detergent.

## Surfactants

▼ Detergents and soaps are members of a chemical family called surfactants (short for surface active agents). Their cleaning power comes from their ability to coat droplets of grease and oil so you can flush them away with the water. Neato.

## 5. Psychedelic Milk

This mesmerising trick will keep you spellbound with dancing swirls of colour. Thanks to the cows and dairy farmers who make it all possible... moo-chas gracias!

Fill a shallow plate with water with full cream milk. Note: Skim milk doesn't work.

Add a few drops of each food colour in four separate places.

Carefully squeeze a big drop of detergent onto the edge of the plate so it runs down into the milk.

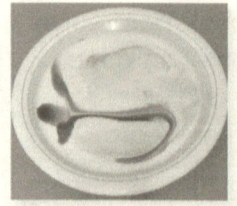

When the detergent hits the milk, amazing things happen. Wait at least ten seconds and a swirling kaleidoscope of colours will appear on the surface! You can add more detergent if it stops.

## What's going on?

Sandologist ON DUTY

You learnt how detergent reduces the surface tension of water in the Pepper Scatter trick on page 16. At first, the same thing happens with milk but the swirling effect is different. It keeps going much longer and starts again when you add more detergent.

Believe it or not, nobody knows exactly how this trick works but here's what we do know. Milk is a very complex mixture of water, protein, a sugar called lactose, tiny fat globules and a myriad of nutrients including calcium. Everything a growing calf needs!

Because this trick doesn't work with skim milk, the fat in full cream milk must be the key. Oil (or fat) and water molecules don't usually mix but detergent molecules can attach to both. This property of detergent is obviously a key factor but for now, exactly what's going on in your plate remains a mystery and I reckon that's pretty cool.

### Homogenisation

▼ After you milk a cow, the cream (fat) slowly floats to the surface. To stop this happening, raw milk is homogenised before you buy it. Homogenise means to mix evenly. Forcing milk through tiny holes under pressure breaks fat globules into droplets just like a garden hose set to fine mist. These droplets don't re-combine so the fat stays evenly mixed in the milk.

# 6 Food Colour Vortex

A drop of food colour in water does an amazing thing. It turns into a swirling donut. Hint: use an eye-dropper for this trick if your food colour isn't in the squeeze bottle type. Be careful with food colour because it can leave stains.

Fill a tall glass to the brim with water and then wait for at least 30 seconds. Even though it may look still, the water keeps swirling for ages!

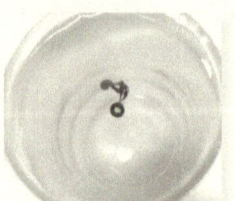

Carefully squeeze a drop of food colour so it is dangling from the eye-dropper. Touch the surface of the water with the drop of food colour and this is what happens...

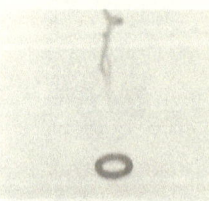

The drop shoots in fast and quickly slows down. As it descends, it turns into an amazing, swirling ring-shaped vortex.

As your vortex descends further, it will break into an upside down crown shape, with more rings on each tip. It's super-cool!

## What's going on?

serious experimentation underway

A swirling fluid is called a vortex. The mathematical name for a donut's shape is a toroid. So you've just made a toroidal vortex. Toroidal vortices form due to a complicated combination of friction and pressure. Friction between the drop of food colour and the water slows its descent. The friction is not equal on all parts of the drop. The 'sides' experience a sideways or shearing friction. The very bottom of the drop meets the water head on and so experiences more friction. The little drop also leaves a little wake of low pressure behind it. The differences in pressure and friction deform the drop into the swirling donut shape.

The sudden expansion and collapse you saw is called a vortex breakdown. These are not yet fully understood but of great interest to scientists who study fluid dynamics.

### Dolphins and Volcanoes

▼ Some volcanoes blow the occasional toroidal vortex too. Those of Mount Etna in Italy can be more than 200 metres wide, last for more than ten minutes and fly one kilometre into the air!

▼ Dolphins have been seen and photographed blowing toroidal air bubbles. It sounds unbelievable but they chase and play games with them, using their pectoral fins to stop them rising too fast. Nice one Flipper!

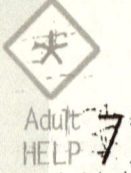
Adult HELP needed

# 7. Balloon on a Stick

This is a classic science trick every kid should know. All you need is a balloon, a shish-kebab skewer (or wooden stick) and a very strong nerve.

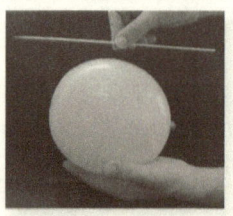

Inflate your balloon so that its diameter is a bit shorter than the length of your skewer.

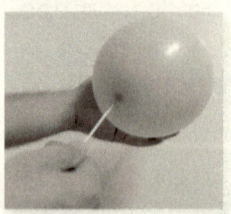

Poke the sharp end of the skewer through the dark spot on the top of the balloon. Be brave and push hard...don't worry, it won't pop!

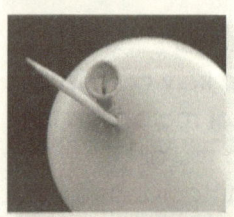

Now gently, push the skewer all the way through to the dark area around the knot and out of the balloon.

Voila! A balloon on a stick!

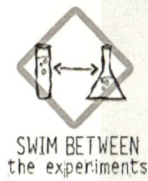

SWIM BETWEEN
the experiments

## What's going on?

Poking a hole in a balloon usually pops it instantly, but you've just poked a skewer through two special parts of the balloon.

When you inflate a balloon, the rubber stretches to more than three times it's normal size! As it stretches, the rubber gets thinner but the spot on top and the area around the knot don't get stretched at all. These parts of the balloon are not stretched so they're thicker and darker.

There's a lot of tension in the stretched rubber of an inflated balloon. It's so tense that even the tiniest tear will grow at lightning speed and pop the balloon.

But a hole won't grow where there's no tension. That's why you can poke a skewer through the darker, thicker parts without your balloon going pop!

## How balloons are made

▼ Balloons are made from a natural substance called latex which is the sap of rubber trees. Latex is incredible stuff. It can be stretched to amazing lengths and it's totally biodegradable. It's naturally clear but pigments can be added to make any colour you like.

▼ Balloons are made by dipping balloon-shaped metal moulds into liquid latex. The moulds are then dipped in a chemical which causes the latex to set. The dark spot at the top of a balloon is a result of the liquid latex flowing down the mould before it sets.

Adult HELP needed

# 8 Fireproof Balloon

This mind-blowing science trick will shock and astound any audience. Use a dark coloured balloon if you want the water inside to remain a secret.

Fill a balloon to the size of an orange with water – I use a sports bottle but the tap works fine too. Then inflate and tie the balloon as usual. I've added food colour so you can see the water.

Now do the unthinkable...gently lower the balloon so that it is touching the candle flame!

Look closely. The flame is touching the balloon but not popping it.

Lift the balloon up, then away from the flame. Rub the black spot with a wet finger and you'll see that it's not burnt rubber, but soot from the candle.

SCIENTISTS
patrol these waters

## What's going on?

With just air inside, the rubber gets too hot and the balloon goes pop! Water is much heavier than air so it can absorb more heat. But this trick won't work with sand, flour or even vegetable oil instead of water, so there's a bit more to it than that.

Like all fluids (gases and liquids), water expands when it heats up. As a result, hot liquid rises just like hot air. So water heated by the candle flame rises, which draws in cooler water from nearby. The cooler water heats up, rises and this cycle continues until all the water reaches 100°C. Balloon rubber has to get hotter than 100°C to pop but the rising water absorbs the heat and the balloon survives.

Candles burn wax very efficiently but if you put something solid in the flame, the vapour condenses before burning completely. The black spot you get is not burnt rubber but partially burnt candle wax vapour called soot.

## Water's amazing heat capacity

▼ Suppose you had one kilogram of steel and one kilogram of water, both at room temperature. If you put them on a stove so they absorbed exactly the same amount of heat, which one would get hotter? Surprisingly, the temperature of steel rises ten times faster than water! It also takes ten times longer for water to cool back down again.

## 9. Leakproof Bag

Do sharp pencils and plastic bags full of water make a recipe for disaster? Your experience with balloons would say yes but the result says no...sort of.

Find a plastic bag that doesn't leak and fill it with water. I added food colour to the water so you can it see more clearly.

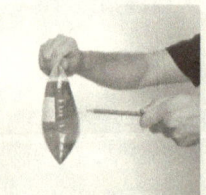

Now do the unthinkable and poke a pencil right through the bag. It won't leak!

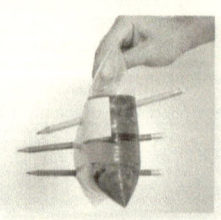

Stab as many pencils through as you like, still no leaks!

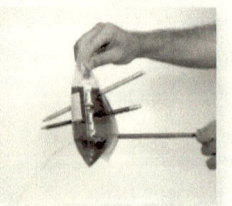

Take a pencil out and your bag does a wee – tee-hee. But put the pencil back in the same hole and it stops leaking. Hey, a toilet trained bag!

## What's going on?

Sandologist ON DUTY

You'd expect a bag full of water to burst open like a balloon but it doesn't. That's because plastic bags are made from a type of plastic called polyethylene. If you've ever tried to tear a plastic bag or cling wrap, you'll know it stretches before it tears. That's half the story of how this trick works. The other half is the shape of your pencil.

The sharp tip of your pencil makes a very small hole in the bag. As you push, the plastic stretches just enough to accommodate the thicker part of your pencil. The result is a perfectly watertight seal.

## Plastic bag problem

▼ Australians use 6.4 billion (6,400,000,000) plastic bags every year. If you need convincing that this is a bad thing, try this experiment. Bury a plastic bag alongside a large leaf in your garden and mark the spot. Set a reminder to dig them both up in four weeks time. You'll find the leaf has been almost completely eaten by insects, worms and bacteria. Together, they will eventually recycle the entire leaf back into the soil. The plastic bag however, will still be intact. That's because nothing eats plastic and it doesn't dissolve or break down. Plastic is not biodegradable and lasts for hundreds of years. Now multiply this small quandary by 6.4 billion and there's your problem!

## 10. Spear a Spud

Tell someone you're going to slam a drinking straw through a raw potato and they'll laugh. But watch their face when you accomplish this amazing feat with ease...ha-ha! Howzat!

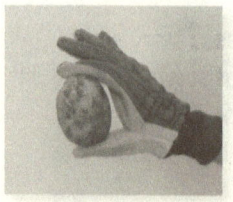

Hold the potato between your thumb and index finger as shown. Make sure you wear a glove for safety; a washing-up glove is fine.

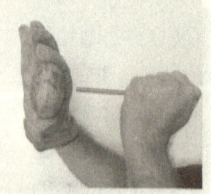

Hold the straw firmly, take aim and smack it into the potato as hard and straight as you can.

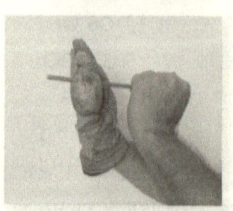

The straw easily slices right through the potato and out the other side.

Pull the straw out and you're left with a hole in your spud. If you're wondering where the missing bit is, look inside your straw!

## What's going on?

The beauty of this trick is that no one expects it to work. Most people think the straw will crumple as though it slammed into a brick wall.

serious experimentation underway

There are two reasons the straw slices through the potato so easily. First, straws are really just long hollow cylinders. They might be made of flimsy plastic, but like all cylinders, straws are amazingly strong under compression. Try squashing one between your index fingers.

Second, straws are made from very thin plastic and that makes them pretty sharp. So held the right way, a straw is sharp and surprisingly strong little dagger that easily cuts through a raw potato.

Stab the potato repeatedly and you can make cylindrical chips! Pretty cool, eh?

## A brief history of straws

▼ Before plastic came along, straws were made from waxed paper. They're very easy to make. Just wind a narrow strip of paper around a pencil at a slight angle. A bit of glue applied along the edge easily holds it together. To stop them absorbing water and going limp, these cylinders were coated in wax and, hey presto, a paper straw.

Adult HELP needed

# 11 Super Can

How many books can you stack on top of an empty soft drink can before it crumples? You will most definitely be amazed!

Use big books with hard covers and stack them so they are centred on the can.

Keep stacking on the books. It soon becomes very, very wobbly so watch your toes!

If you run out of books before the can gets crushed, tap it with a stick. The slightest deformation will bring your stack down with a loud crunch.

I managed to stack 52 kilograms onto this can before it collapsed under the weight... flat as a pancake!

SWIM BETWEEN
the experiments

### What's going on?

This trick demonstrates the amazing strength of cylinders. Aluminium soft drink cans are amazingly strong considering each one weighs just 15 grams. Their walls are only 0.11mm thick, which is about the diameter of an average human hair.

Aluminium cans are very easy to crush when you lay them flat. But like all cylinders, they're surprisingly strong when loaded the right way. Standing upright, an empty can will support more than 53 kilograms. An unopened can full of soft drink can support more than 100 kilograms.

Cylinders are strong under compression because they distribute their load very evenly. But even the slightest defect can dramatically reduce a cylinder's strength. Under a heavy load, the slightest dint can cause a cylinder to collapse.

### Recycling saves

▼ It's not just the aluminium you're saving when you recycle. The energy required to make one new can is the same amount it takes to recycle it 20 times. Australians use more than 3 billion aluminium cans per year so there's a lot of energy to be saved by recycling too!

## 12 Uncanny Cancan

Turn an empty soft drink can into a groovy ballerina that really can do the amazing aluminium can cancan.

Pour about 100 ml of water into an empty soft drink can so that it's roughly a quarter full.

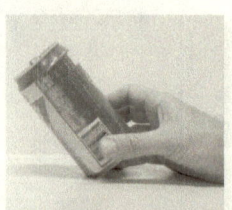

Tip the can onto its rim at an angle as shown and you'll find it stays perfectly balanced.

A can balanced like this attracts a lot of attention!

The can *can* balance because of the shape of the rim. Give the can a gentle sideways nudge and it will even pirouette like an aluminium ballerina.

SCIENTISTS patrol these waters

## What's going on?

For an object (or animal or person) to balance, its centre of gravity must be directly above its point of support. If you lean too far forward, your centre of gravity will no longer be above your feet and you fall forward. The centre of gravity is an average position for all the weight of an object.

If you try to balance a full or empty can in the same way, it will fall over as expected. That's because the centre of gravity for a full or empty can is roughly in the centre of the can.

Aluminium cans weigh 15 grams and 100 ml of water weighs 100 grams. On an angle, the centre of gravity with this much water is directly above rim. And because water is a liquid, it always ends up being level inside the can. The can is able to pirouette because the water can flow to retain its shape.

## Ballast water

▼ Seawater is used to weigh down unladen cargo ships for stability. This shifts the centre of gravity toward the keel so the boat doesn't capsize. The problem is that seawater contains lots of organisms which will travel with the ship. If the ship empties that water in another country, those organisms can become a pest. More than 100 exotic species are known to have entered Australian waters this way so strict rules have been introduced for transferring ballast water.

Adult HELP needed

# 13 Straw Atomiser

This trick makes a cool whistling noise and it's messy which means it *must* be cool! It might help you teach adults how those antique insecticide sprayers work too.

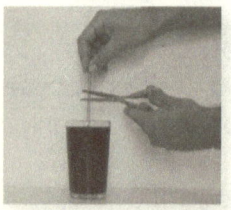

Fill a glass almost to the brim with water. Then trim a straw so it is a few centimetres taller than the rim of the glass.

Put a second straw in your mouth and hold it perpendicular to the one inside the glass. I added food colour so you can see the water clearly.

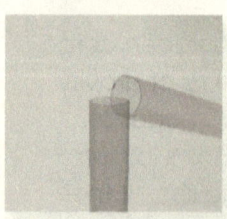

The horizontal straw should point across the top of the one in the glass like this.

Blow hard and this is what happens. Aaaaargh!

## What's going on?

Sandologist ON DUTY

This trick proves a groovy but slightly confusing fact about fluids (gases and liquids). Inside a still fluid, the pressure is equal in every direction. But when fluids flow, they simultaneously exert more pressure in some directions, and much less in others.

When you blow through a straw, a stream of air rushes out. Aim that stream directly at an object like a plastic cup, and it goes flying. But aim along a surface and the opposite happens. At right angles to the very same stream, the pressure is lower.

Normally, the pressure at the top and bottom of a straw resting in a glass of water is equal. But when you blow across this straw, the pressure at the top is reduced. Now there is a pressure difference between the top and the bottom. Fluids always flow from high to low pressure so water flows up the straw towards the lower pressure. When it reaches the top, the same stream of air blasts the water sideways to produce a fine mist. Sweet!

## How Atomisers Work

▼ Perfume atomisers and old-fashioned insecticide sprayers work on the same principle. A pump blows air across a vertical tube (like a straw) sitting in a reservoir of liquid such as perfume or insecticide. The liquid rises to the top of the tube where it gets hit by the stream of air. Result: a fine mist of perfume or insecticide.

# 14 Levitation Trick

This trick will make a ping-pong ball, marble or balloon appear to magically levitate in midair! All you need is a bendy straw, a spherical object and a healthy set of lungs.

This works with ping-pong balls, air inflated water bombs, marbles, round chocolates or mints...anything that's small and spherical.

It's not necessary, but why not draw a smiley face on your ping-pong ball?

Rest the ball, marble or balloon on the end of the straw. Keeping it balanced is tricky.

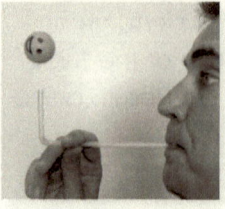

Blow as long and hard as you can. Instead of falling to the ground, the ball hovers directly above the straw. Cowabunga little dude! Weeeeeee!

## What's going on?

serious experimentation underway

It's that weird pressure-reducing habit of flowing fluids at it again. The stream of air you blow through a straw comes out pretty fast. Where it belts into the bottom of the ball it pushes up. But the air keeps flowing up further and *around* the ball, which does a very cool thing.

When air flows along a surface it reduces the pressure at right angles to it. The air flowing up all around the ball exerts less sideward pressure than still air and this keeps it firmly stuck in the stream.

You might have seen the same trick done with a beach ball in vacuum cleaner shop windows. Most vacuum cleaners can blow air as well as suck air and the plastic seams cause the beach ball to spin. This eye-catching display lures unsuspecting customers into the shop where sales staff are waiting to pounce. Bam! Another sucker sold.

## Bernoulli's Principle

▼ The effect that makes this trick work is named after the Swiss scientist and mathematician Daniel Bernoulli. He derived the famous equations that accurately predict the pressures inside flowing fluids.

▼ Now you'd think Daniel's dad, also a mathematician, would have been pleased as punch. But Papa Bernoulli's ego was not well equipped for a high-flying son. After winning a joint prize for work they'd done together, Daniel's dad got angry and banned him from the family home. They hardly spoke again. Nice one Papa Bernoulli...ya dingaling!

# 15 Sticky Stream

Use the weird and wonderful laws of nature to stick a ball to a flowing stream of water. Sounds too strange to be true...

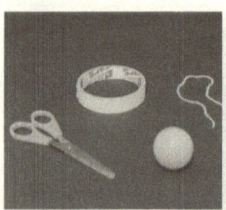

You need a ping-pong ball, some strong tape, string and scissors.

Cut a small square of tape and use it to stick the string firmly on the ball.

Hang the ball in a stream of running water and slowly pull the string away.

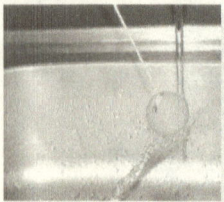

Hey wowsers! The ball is stuck to the stream. It feels even weirder than it looks. Please do this with a plug in the sink and recycle the water wisely.

## What's going on?

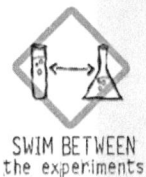

SWIM BETWEEN
the experiments

This amazing trick demonstrates how streams of fluid stick to a curved surface. It is named the Coanda Effect after Henri Coanda who is widely regarded as the godfather of modern jet planes.

Gravity pulls everything straight down but as water flows around a ball, it changes direction and pours off at an angle. Now that famous fact of nature called Newton's Third Law joins in. Every force is opposed by an equal and opposite force. The force of the water pouring off the ball in one direction generates an equal and opposite force in the other direction. This keeps the ball stuck firmly in the stream. It looks and feels weird, but it's all ship-shape and above board in the surprising world of physics.

### Coanda v Bernoulli Controversy

▼ If you ever talk to an aeroplane enthusiast about airflow over wings, be warned. There's still a lot of confusion over how wings generate lift. Some people claim it's Bernoulli's Principle that keeps planes aloft. Others will argue it's the Coanda Effect. You'll find websites arguing fiercely in both directions. Even scientists can get into wonderfully heated arguments over it. To be honest, it confuses me too but I get a real kick out of watching this turbulent debate. Baboom! Airflow and turbulence. Get it?...hmmm.

# 16 Inseparable Books

This will blow your science teacher's mind. Interleave the pages of two books and you'll discover it's impossible to pull them apart. No glue, no staples, no tricks!

Find two identical or very similar books with at least 100 pages each.

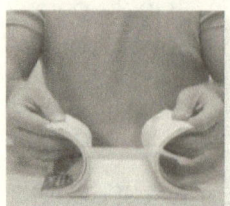

Carefully interleave the pages like a deck of cards – the more evenly you do this, the better the trick will work.

The books should overlap to about the middle of their pages as shown in the picture.

Hold each book by its spine and try to separate them – it's impossible! Amazing huh?

SCIENTISTS
patrol these waters

## What's going on?

It's nothing more than plain old friction that prevents you from pulling the books apart. Friction is the name we give to the force that opposes the motion of two surfaces in contact. It's barely noticeable between two sheets of paper. But multiply this relatively small amount of friction by a few hundred pages and it adds up to be insurmountable.

The friction between two surfaces depends not only on the materials they are made of but also on how rough or smooth they are. Two planks of wood will slide over each other more easily if you make them smooth with sandpaper. But even very smooth surfaces have some friction between them. Adding a fluid such as oil or air between two surfaces can reduce friction. Fluids used to reduce friction are called lubricants.

## A fraction too much friction trivia

▼ The study of friction is called tribology and it's more widespread than you might think. Ball bearings, tyres and automotive lubricants are obvious examples but lipstick, hair conditioners and artificial hips also rely on the study and measurement of friction. So there you go – a cool science trick, some friction trivia and a weird new word to add to your smarty-pants vocabulary. That should help you rub a few people up the wrong way... shazam!

## 17 Gripping Rice

Challenge a friend to lift a jar of rice using nothing but an ordinary pencil. Make 'tut tut' sounds while they're failing miserably, then show them how easy it is to do.

Fill an empty jar to the brim with uncooked rice. The jar's mouth must be narrower than its sides. (A drinking glass won't work.)

Jab a pencil deep into the rice repeatedly, sharpened end down. You may need to do this more than 40 times.

The rice will settle and make a little crater around your pencil. Add a little more rice to top up and keep jabbing – be patient, this really works.

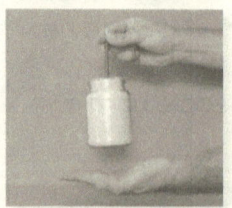

When it gets really difficult, push the pencil in deep and carefully lift the jar. Keep your hand under the jar and don't lift too high. To release, just twist and pull the pencil out.

### What's going on?

Sandologist
ON DUTY

The shape of the jar and the tip of your pencil make this trick work. It won't work with the blunt end of your pencil or a normal drinking glass.

Initially, the grains of rice in a jar are fairly loosely packed together and there is little friction between them. The tip of your pencil acts like little a wedge which pushes the rice sideways. Every jab compacts the rice a little more and some even gets forced up which makes a little crater. The narrow mouth keeps the rice in the jar. It doesn't work with a normal drinking glass because rice will spill out over the rim.

After forty jabs, the grains are packed in so tight that it becomes very difficult to push the pencil in. Eventually, the friction between all the grains of rice, the jar and the pencil is so large that you can lift up the jar.

### Friction ... Yin & Yang

▼ When you're surfing, you want lots of friction between your feet and the board, which is why surfboards need wax. Otherwise you slip off and wipe out. Friction is good!

▼ When you're skateboarding, you want very little friction between your axle and wheels which is why you need bearings. Otherwise you won't get enough speed and will miss the ramp. Friction, in this case, is bad!

## 18 Mini Icebergs

It's amazing how mesmerising really simple things can be. Make some coloured ice cubes, float them in a glass of water and watch the spectacle unfold.

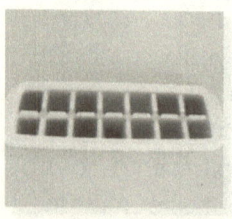

Fill an ice cube tray with water. Add four or five drops of food colour to each section and stir. Be careful with food colour as it can stain.

The first thing to notice once your cubes are frozen is that all the colour gets concentrated into the middle of the ice cubes.

Use tongs to add one coloured ice cube into a glass of water. Be gentle to keep the water still.

Watch carefully. You'll observe at least three amazing things happening.

## What's going on?

Amazing things happen when water freezes and melts. As water freezes, the food colour gets concentrated in the centre of the cubes. That's because pure water freezes first. Any impurities get squished to the middle of the cube. That includes dissolved gases which is what the bubbles you see in normal ice cubes are.

When you float an ice cube in water, you're witnessing another amazing thing. Solid chunks of nearly every other natural substance sink in their liquid form but ice does the opposite. That's because most substances shrink as they freeze but water expands! This is just one of water's many bizarre properties.

Streams of colour descend from the ice cube because water shrinks as it melts. The streams slowly rise again which eventually mixes all the colour evenly. This process of descending and rising due to different temperatures in a fluid distributes heat and is called convection.

serious experimentation underway

## Iceberg rollover

▼ If you're lucky, you might even see your ice cube roll over. This happens when the bottom melts enough for the cube to become top heavy. Real icebergs sometimes roll over spontaneously for the same reason. You have to be very lucky to witness this but it's one of the ocean's most awesome spectacles.

*Adult HELP needed*

# 19. Ice Cube Lava Lamp

This trick turns the traditional lava lamp on its head. Float a coloured ice cube in cooking oil and watch blobs of water sink and erupt in bursts of colour.

Quarter fill a glass with water. Tip the glass on an angle and gently pour cooking oil down the side until full.

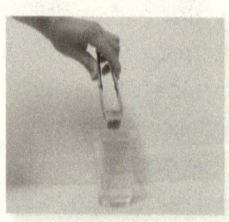

Use tongs to gently lower a coloured ice cube into the oil. Notice how liquid water sinks in oil but ice cubes float.

Watch carefully as the ice melts. Drops of water form and tip the ice cube on a slight angle before sinking into the oil.

The drops take a while to penetrate into the water below but when they do, they erupt in a burst of colour.

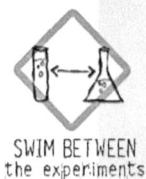

SWIM BETWEEN
the experiments

## What's going on?

Oil and water don't mix but stay totally separate with the oil floating on top. But while oil floats on liquid water, frozen water floats on oil. What a brain twister!

A substance will float on water if it is less dense. One litre of water weighs 1000 grams (1 kilogram). One litre of oil only weighs about 910 grams so oil floats on water.

But as water freezes it does a very unusual thing; it expands! Expanding makes a substance less dense so liquid water sinks in oil, but frozen water floats. When ice melts it shrinks again. That's why melted liquid water drips off the ice cube and sinks into the oil. It's a like lava lamp but going in reverse.

## Ice ain't ice

▼ The molecules in crystalline solids are arranged in neat patterns. Normal ice has a hexagonal crystal structure, which is at the heart of every snowflake. But if water is cooled down really fast, amorphous ice forms instead. The molecules in an amorphous solid are arranged randomly and do not have a crystal structure. Glass is a good example of an amorphous solid. Amorphous ice doesn't float in water but you need special equipment to make it. Many other types of weird crystalline ice can be made in the lab but personally, I prefer water in its liquid state...it's the only form you can surf!

## 20 Ice on a String

This nifty trick will keep you amused at the dinner table if the conversation dries up. All you need is a glass of water with ice, string and salt.

Float some ice cubes in a glass of water.

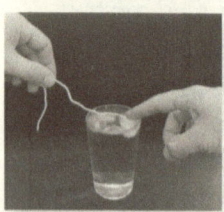

Wet a piece of cotton string and rest it on top of the ice cubes.

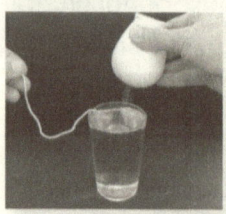

Sprinkle salt on the string and wait about five seconds.

Lift the string and the ice cubes come with it.

SCIENTISTS patrol these waters

## What's going on?

Fresh water freezes at zero degrees celsius (0°C). Seawater freezes at −1.8°C. That's because salt lowers the freezing temperature of water below zero.

When you pour salt onto an ice cube, the ice on top melts. But the salty water that forms quickly sinks into the fresh water and becomes diluted (less salty). As it becomes less salty, the freezing temperature rises back up towards 0°C and some of the water refreezes. When it does, the string gets stuck to the ice cube and you can lift it right out of the water. Cool!

## Pack ice

▼ When seawater freezes, salt gets trapped in the ice. The trapped salt gets concentrated into tiny pockets called brine cells. Seawater is about 35 parts per thousand salt (two teaspoons of salt for every glass of seawater). Brine is more than 50 parts per thousand salt.

▼ Brine cells slowly melt the ice below them so they gradually migrate down. The water above the brine cells is less salty so it refreezes. This very gradual process is called brine rejection.

▼ Sea salt contains several kinds of salt – not just the sodium chloride you put on your chips (sparingly of course!). It also contains minuscule amounts of all the elements that occur in the Earth's crust including gold.

# 21 Flying Cups

This kooky little trick will keep you amused at those intense family barbecues. Your great aunt will marvel at the dizzying altitudes your flying cups achieve!

You will need two empty plastic cups.

Put one inside the other.

Hold them close to your mouth and blow hard between the rims.

Phoomba! The top cup pops out and flies up, up, and away!

## What's going on?

Sandologist
ON DUTY

As you discovered earlier, fast moving air flowing over an object exerts lower pressure (see tricks 13 and 14). The phenomenon is called Bernoulli's Principle.

When you blow between their rims, air flows around the cups really fast. This moving air creates an area of low pressure between the rims. The air trapped between the bottom of the cups remains at normal atmospheric pressure. So the higher pressure inside the bottom cup pushes the top one up.

If you blow gently, the top cup rises slowly and stays in the cup. If you blow hard enough, it shoots up and gets blasted away from you.

## Weird boat

▼ In the 1920s, a chap called Anton Flettner proposed a very odd-looking ship. It used the force generated by air flowing around very large cylinders for propulsion. The cylinders had to be rotated by an engine to make the boat move. Anton was hoping these rotating cylinders would replace masts and cloth sails. Amazingly, this odd contraption actually worked...sort of. The problem was it produced less forward propulsion than you get by simply using the motors to run a standard propeller. Ten points for effort though!

Adult HELP needed

# 22 Flying Tea Bags

Turn an ordinary tea bag into a tiny thermal rider that soars to incredible heights. You'll need an adult helper and a very still room for this trick.

Take the staple out of a tea bag. Have a wet tea towel handy for mishaps.

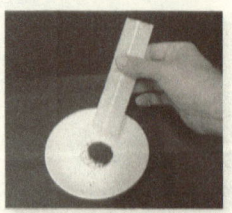

Unfold the tea bag and discard the tea inside. You can still use it to make tea in a pot, otherwise, chuck it in your compost bin.

Open the tea bag into a tall cylinder and stand it on a plate or saucer. The slightest breeze will knock it over so hold your breath and carefully set it alight.

Before the flame reaches the plate, the tea bag takes off. The flame goes out quickly but the ash keeps soaring all the way to the ceiling! Cool.

## What's going on?

serious experimentation underway

Tea bags are made from soft, lightweight paper. When you light it, the flame produces heat. You can't see it, but the hot air produced rises all the way to the ceiling.

A burning tea bag also gets much lighter and soon becomes light enough to hitch a ride on the rising hot air.

The higher your ceiling and the stiller the air the better. I've seen them reach five metres inside a lecture theatre! The ash descends slowly giving you plenty of time to get underneath for a perfect, ultra-slow motion catch. Howzat?

## Thermal Riders in the Sky

▼ When the Sun heats up land, air near the ground starts to rise. Cool air moves in from nearby to replace the rising air but this too gets heated and rises. This process continues as long as the sun keeps beating down.

▼ Columns of rising hot air are called thermals. Many birds like the Australian Wedge-tailed Eagle are experts at finding them. They use them to soar high into the sky without flapping their wings. From there, they can spot rabbits more than one kilometre away and reach speeds in excess of 100 km per hour during a dive. It must be cool being a Wedge-tailed Eagle, although not so cool for the rabbits ...

Adult HELP needed

## 23 Marble Gravitron

Here's a cool way to pick up a marble using a glass and the laws of physics. And if you've ever been on that show ride called 'The Gravitron', you'll know how the marble feels.

You need a glass that's narrower at the top like the one on the left. The one on the right won't work because it's wider at the top.

Cover the marble with your glass on a smooth table. Wiggle the glass back and forth so the marble starts zooming around inside.

Smaller, faster wiggles make the marble spin around faster. The marble is a faint blur in this photo because it's going really fast.

When the marble is really hurtling, raise the glass and the marble comes with it. If you can keep wiggling in midair at just the right speed, the marble stays inside the glass even longer.

SWIM BETWEEN
the experiments

## What's going on?

A centripetal force keeps your marble stuck to the glass. You experience the same force pulling you sideways every time your car goes through a bend in the road or around a corner.

The magnitude of the centripetal force on a moving object depends on its mass, speed and the radius of the curve. The faster you go, or the sharper the bend, the stronger this force will be.

But the centripetal force only pushes your marble sideways, not up. That's why this trick won't work in a straight glass. The marble only rolls up the walls if the glass is the right shape (ie. narrower at the top). As long as the centripetal force is strong enough to stop it rolling back down, the marble stays in the glass.

## Centrifuge

▼ Washing machines use the centripetal force to dry clothes during the spin cycle. Water trapped in your clothes experiences a strong centripetal force and flows sideways. Small holes in the spinning bowl allow the water to escape into the basin and down the drain. A super yet natural being then hangs your clothes out to dry, folds and returns them as if by magic to your room. Thanks Mum!

# 24 Balloon of Death

Have you ever seen that stunt where motorbikes go upside down inside a spherical cage? This is exactly the same, but completely different. Try it and you'll see what I mean.

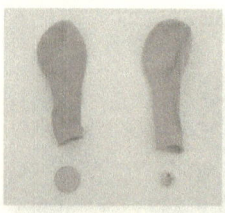

Grab a five cent coin and if you can find one, a hexagonal nut. Make sure there are no sharp edges on either.

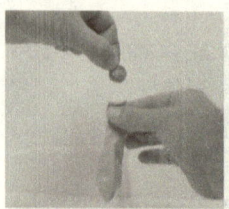

Pop your coin inside a light-coloured balloon so you can see the coin inside. Inflate the balloon but don't tie a knot so you can get the coin out again.

Shake the balloon. The coin flips up on its rim and then rolls around and around at a million miles an hour (I'm exaggerating but it goes really fast!).

Now try it with your hexagonal nut. It does the same thing but also makes a really groovy and slightly annoying zzzooooom noise.

SCIENTISTS patrol these waters

## What's going on?

It's a nifty fact of nature that a tumbling coin tends to flip up and roll on its edge. The reason lies in a property called the moment of inertia.

Every piece of matter in the universe has inertia. It's the measure of how easy or difficult it is to make an object move. The heavier an object, the greater its inertia and the harder it is to get it moving – simple! But spinning is a unique type of motion. Objects of equal weight can be easier to spin than others depending on their shape. The measure of how easily different shapes spin is called the moment of inertia.

Now a disk can spin like a rolling wheel, or tumble like a flipped coin but here's the thing. A disk's moment of inertia is very different for these two types of spin. One object, two moments of inertia. Spinning has the lower moment of inertia so it's easier to spin a disk than to tumble it. That's why a tumbling coin tends to end up rolling on its edge. A hexagonal nut behaves the same way but vibrates the balloon as it rolls which makes a cool zooming noise. It's your very own stunt nut riding the balloon of death!

## The real thing

▼ Riding a motorbike inside a spherical cage obviously takes skill and nerves of steel. But how fast do you have to go to avoid going splat? Surprisingly, not very fast. The cage is typically around 4 metres in diameter. To overcome gravity while upside down, the rider only needs to maintain speeds around 25 kilometres per hour.

# 25 Flippin' Coin Trick

This trick looks like magic. It's very simple but takes a bit of practice. Once you're an expert, everyone will want to know how on Earth you do it!

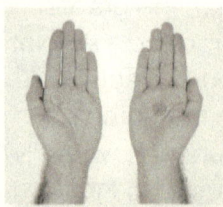

Two-dollar coins work best for this trick. As shown in the picture, the coin in my left hand is near my thumb. The one on the right is in the middle of my palm.

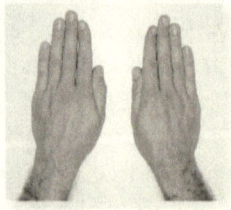

Simultaneously flip both hands and slap them down onto the table. The coin in your right hand will fall straight down. The coin in your left hand will flick across and under your right hand.

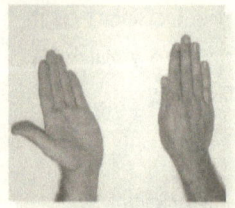

Ask your audience where the coins are. They'll guess there's one under each hand.

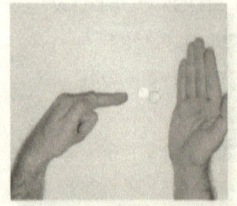

Now reveal both coins under your right hand. Ta-da!
Now, can you make the coins land under your left hand instead?

## What's going on?

Sandologist ON DUTY

Carefully positioning the coins in the palms of your hands makes this trick work. When you flip your hands from 'palms up' to 'palms down', a coin in the centre of your palm will fall straight down as expected. But a coin near your thumb gets flicked across to the other hand.

It's easier to understand if you imagine using table tennis bats instead of your hands. Imagine laying a coin in the centre of one bat. As you rotate the bat, the coin just falls to the table. But lay a coin near the edge of the bat and the coin gets flicked across the table.

With a bit of practice, it all happens so fast that an observer won't notice this before you slap your hands down.

You can add a little pizazz by pretending to pick up the 'missing' coin and pretending to push it through the top of the hand with both coins under it. This looks freaky!

## Coin trivia

▼ Did you know there's a limit to how many coins you can use to pay for a single purchase? The Australian Currency Act stipulates that you can't use more than $5 worth of 5c, 10c, 20c or 50c coins at a time. Bizarro!

▼ The rarest Australian coin is the 1930 penny. Only six are known to exist so if you happen to find one you'll be very rich!

# 26 Science on a Shoestring

Tie some keys and a ring to a shoelace, hang them over a pencil and let go. Will they hit the floor? Try it and discover the surprising result for yourself.

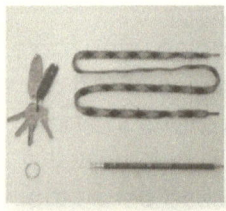

Grab a bunch of keys, a pencil, a shoelace and a ring.

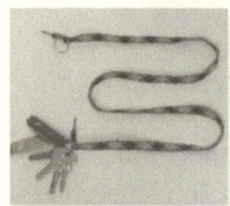

Tie the keys to one end of the shoelace, and the ring to the other.

Hold the ring and hang the shoelace over a pencil as shown.

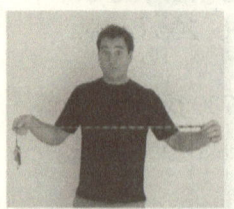

Let go of the ring... it starts to fall but so do the keys. The ring then zips around the pencil three or four times, gets stuck and stops the keys falling further.

## What's going on?

serious experimentation underway

This tricky trick is easy to perform, trickier to predict but really tricky to explain.

The keys are much heavier so they fall straight down, which pulls the ring up. The ring starts to fall like a pendulum but the length of shoelace to swing on gets shorter by the millisecond. When it reaches the other side, the ring is going so fast that it overshoots the height you released it from. As the ring zips around the pencil, friction between the loops of the shoelace slows the keys down to a complete stop.

The motion of a pendulum such as a ring hanging on a string is called simple harmonic motion. Harmonic refers to the regularity of a pendulum's swing. By studying simple harmonic motion, scientists can learn about the motion of the tiniest atoms right through to planets and solar systems. The motion of the ring in this trick is much more complicated though. It would take several pages of mathematical formulae to describe!

## Flugelbinders

▼ Have you ever wondered what the hard plastic or metal tips on the ends of a shoelace are called? The character played by Tom Cruise in *Cocktail* called them 'flugelbinders'. Thanks for the tip, Tom, but the real answer is *aglets*, from the French word *aiguillette*, which means 'small needle'. Next!

Adult HELP needed

# 27 Anti-gravity Coaster

Have you got the guts to turn this glass of water upside down and let go of the coaster? Well of course you have, but try it outside because it can go disastrously wrong.

Fill a plastic cup or glass with water. It doesn't have to be completely full.

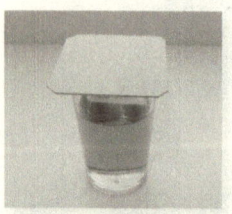

Cover the glass with a coaster or piece of smooth flat card. Smooth is the key word here. Don't try it if the coaster is bent, frayed, soggy or buckled.

Hold the coaster and turn the glass upside down. You're ready to do the unthinkable. Let go of the coaster.

Hey, it's stuck. No leaks? Important: make sure you hold the coaster when you turn the glass over again.

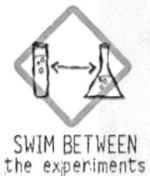

SWIM BETWEEN
the experiments

## What's going on?

Atmospheric pressure and the surface tension of water combine to make this trick possible.

For water to pour out of a glass, an equal volume of air has to get in. Without the coaster, there's no problem and water just pours out. But water sticks to itself, the coaster and the walls of the glass so tight that no air can sneak in. The ability of a liquid to stick to itself results in surface tension and water has this ability to an exceptionally large degree.

But there's another reason the coaster gets stuck to the glass. Air pressure is the weight of all the air in our atmosphere being pulled down by the Earth's gravity. We rarely notice this pressure because it is equal in every direction. Air flows so easily that we can walk around in it with ease. But this barely noticeable pressure pushes up on the coaster so hard that it stays firmly stuck to the glass.

It doesn't take much of a bump to dislodge the coaster so do this outside or over the sink.

## Coasterology

▼ Coasters absorb condensation which keeps tables nice and dry. But as is often the case with cheap, mass-produced artefacts, collecting coasters can become an obsession. It's called tegestology from the Latin *tegetis*, meaning covering or mat. A tegestologist is a person who collects and cherishes coasters from cafes, restaurants and hotels wherever they go. I'm not kidding but don't ask me why they do it!

# 28 Magic Hanky

Apart from its primary function related to the human nose, the humble handkerchief is capable of some astonishing feats. Here is just one of it's many other hidden talents.

Fill a wine glass with water and cover with a handkerchief. I've added food colour to make the water visible.

Push the hanky into the glass with your finger so that it gets wet.

Now pull the sides of the hanky back down the side of the glass so that it is as taut as the skin on a drum.

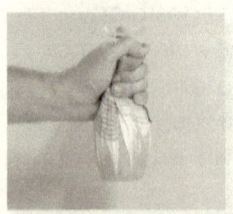

Hold the hanky tight around the stem of the glass. Now do something crazy... turn the glass upside down! Amazingly, the water stays in the glass... ta da!

SCIENTISTS patrol these waters

## What's going on?

It's atmospheric pressure and water's amazing surface tension at work once again.

The explanation is almost identical to the Anti-gravity Coaster trick. But the big surprise with this trick is that no water leaks through the hanky. So, why not?

We all know that handkerchiefs are not very waterproof. The fibres quickly get wet and water can pass right through the tiny spaces between them. But for water to pour out of a glass, an equal volume of air has to get in. To do that, air has to overcome the strong surface tension of water between each fibre in the hanky. The air pressure outside the glass pushes up against the water in the hanky so hard that no water leaks out.

A hanky made from loosely woven cloth might leak one or two drops as you turn the glass upside down. More finely woven materials work better and most won't leak a drop.

## Hanky folding

▼ Handkerchiefs can be raised to signal surrender in battle, worn on the head as a ridiculous hat, or tucked into jackets to look stylish and suave. There are at least 12 ways to fold a handkerchief worn in the breast pocket of a suit. Of these, the Presidential is the simplest. But there is also the One-point fold, plus the Two, Three and Four-point folds. Then there's the TV fold, the Cagney, the Puff and Reverse Puff, the Astaire, the Straight Shell and the Diagonal Shell. I'll just stick with T-shirts thanks.

## 29. Polarised Sunnies Trick 1

Ever wondered what the big deal is with polarised sunglasses? Grab a pair and any gizmo with an LCD screen, such as a laptop or mobile phone, and find out.

Put the polarised sunnies on and look at your LCD screen. I'm holding a pair in front of my laptop so you can see what happens.

Slowly rotate your gizmo (I've rotated the sunglasses instead). Can you see what's happening? No light is getting through the lenses. Keep rotating and light comes through again.

Now you know what to look for, go outside and look at the horizon or the sunlight reflecting off windscreens and bitumen roads.

The horizon and reflected sunlight don't get blocked out completely, but they get heaps darker.

## What's going on?

Sandologist ON DUTY

Light is a type of wave called an electromagnetic wave. But to understand what's happening with your sunnies, let's think about a different type of wave first.

If you wiggle a long piece of rope, waves travel along it. If you wiggle the rope vertically (up and down), these waves are vertically polarised. If you wiggle the rope horizontally (side to side), the waves are horizontally polarised.

Now imagine that you've threaded your rope through a picket fence or swimming pool fence. If you wiggle the rope up and down, the waves can pass straight through the fence uninterrupted. But if you wiggle the rope side to side, the waves hit the pickets and get completely blocked. You could try this experiment if it's not making any sense. The fence allows vertically polarised rope waves to pass through but blocks horizontally polarised waves. This is similar to light passing through polarised sunglasses. If you align the polarisation of your light and sunglasses, the light gets through. But if you rotate the light or your sunglasses, the light gets blocked.

The Sun emits trillions of individual light waves per second but each one is polarised differently. So collectively, sunlight is unpolarised. Light waves from LCD screens however, are all polarised in the same direction. That's why your sunnies can block all of them out so the screen looks black. And here's the big deal about polarised sunnies. When unpolarised sunlight is reflected at low angles, it becomes polarised. So wearing polarised sunnies blocks out reflected light. Nifty, eh?

# 30 Polarised Sunnies Trick 2

Here's an even more amazing thing to try with your polarised sunnies. Grab some sticky tape and discover the colourful world of circularly-polarised light.

Stick several pieces of plain sticky tape on an LCD screen. Make a star pattern.
Hint: it's easier to get the tape off again if you double the ends over first.

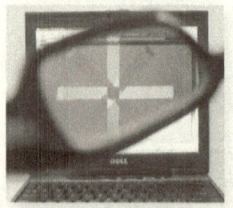

Look at the screen through your polarised sunnies. You'll see colours that weren't there before!
Note: This trick will work best if you display a white image on the screen.

Rotate the sunnies and the colours and brightness will change.

Look closely and you'll see that a really cool star has appeared in my tape.

## What's going on?

White light is a combination of all the colours of the rainbow. Each colour of light is an electromagnetic wave with a slightly different wavelength. A wavelength is the distance between any two consecutive peaks or valleys in a wave.

serious experimentation underway

When light waves travel through some materials like sticky tape, the angle of polarisation rotates. The rope equivalent, (see trick 29), would be wiggling the rope up and down to make vertically polarised waves. But as it travels down the rope, the vertical wiggle rotates to become a left to right waggle. This doesn't happen in ropes of course, but it's a good analogy of what's happening when light travels through sticky tape.

The thicker the sticky tape, the more the light gets rotated. That's why you see different colours where the tape overlaps. But in case you're wondering, you won't learn about circularly polarised light until the third year of a university physics degree. So don't panic if this all sounds super complicated because frankly, it is!

## Brain twister

▼ Light waves aren't easy to visualise like the waves on a rope. They pass through empty space at 300 million metres per second. When light enters your eyes it passes through a lens and gets focused onto your retina. Special cells convert the light energy to a tiny electrical pulse. Your brain processes trillions of these pulses into an image every second. So you see the world but you cannot see the individual light waves that help you see it! It really is bizarre stuff.

# 31 Phantom Patterns

Have you ever seen something that wasn't really there? Well, these phantoms may not be as spooky as a good ghost story, but at least they appear on command!

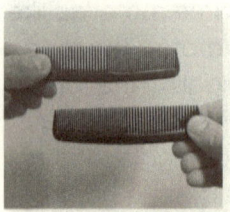

View two combs against a plain white background, like a wall or a sheet of paper. Whoopy-do... nothing happens.

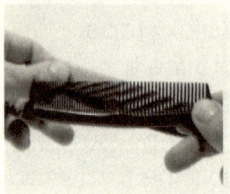

Now hold one of the combs in front of the other like this. Do you see those dark fringes? They weren't there before!

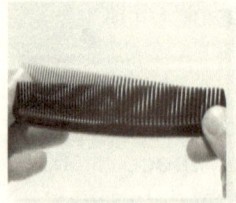

Try moving one of the combs back and forth, or left and right, or rotate it slightly and the dark fringes change. The phantom pattern is on the move.

This is what happens when two identical dot patterns overlay.

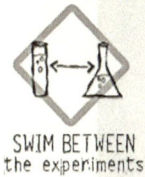

SWIM BETWEEN
the experiments

## What's going on?

These phantom fringes are called Moire patterns (pronounced *more-ay*). They are produced whenever one uniform pattern is laid over another.

The word *moire* is actually a name for fabrics like silk or rayon, which have a wavy or rippled surface. Now that you know what they are, you'll notice moire patterns all over the place. You can see them in the vertical pickets of two parallel fences and many other structures viewed from afar.

Moire patterns can be produced accidentally by scanning printed images too. Look at a newspaper photo closely and you'll see it's actually comprised of thousands of dots.

Scanners sample individual dots, not an entire image but the dots sampled by your scanner never align perfectly with those on the photo. The result is a phantom pattern in your scanned image like the one in the fourth picture on the opposite page. Fortunately, most scanners have a 'descreen' option which gets rid of this problem.

## Chicken combs

▼ Chickens have combs too, but not like the ones in this trick. The domestic chicken's scientific name is *Gallus gallus*. They're members of the *Phasianidae* family of birds which all have combs and includes pheasants, turkeys and partridges. The study of birds is called ornithology and so far, ornithologists have identified about 10,000 species of birds. Maybe that's why they're always so chirpy? Yes, that's a fowl joke!

# 32 Paperclip Linker

If you set this trick up in secret, your friend (or brother or sister) won't be able to figure out how to do it... and that'll annoy them!

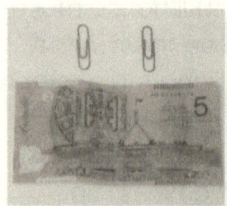

You need two paperclips and a piece of paper or a note.

Fold the note into a Z shape and secure it with the paperclips as shown. (Push the paperclips right down.)

Now pull the ends of the note apart. The paperclips will slide towards each other. Keep pulling and the clips will fling off.

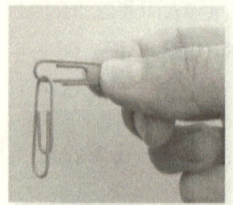

When you pick them up they'll be linked together like this.

SCIENTISTS
patrol these waters

## What's going on?

This simple little trick has a pretty surprising outcome. You expect the paperclips to lock together and get stuck. But if you do it slowly, you can see how it works.

You can add an extra layer of curiosity to this trick with two large rubber bands. Attach the rubber bands first, then the paperclips as illustrated below. What do you think will happen when you repeat the trick this time?

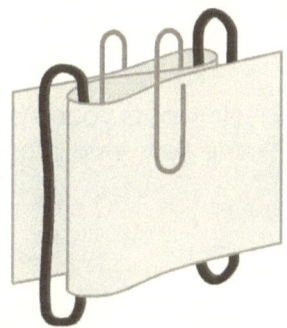

## Paperclip History

▼ Before paperclips, people tied sheets of paper together with ribbons. By the late 1800s, engineers were getting very good at mass-producing bent wire. Dozens of paperclip designs flourished and went into production. The best-known design that we still see today is the Gem paperclip (the one used in this trick). Gem was the name of the company that made them.

# 33 Orange Life Jacket

Legend has it that Sir Isaac Newton nutted out his theory of gravity after being hit on the head by a falling apple. Well, here's another fruity trick that might help with your theory of buoyancy.

Plonk an orange in a container full of water and it floats.

Now peel your orange carefully so the flesh is still in one piece.

Put the flesh of the orange back in the water and now it sinks.

The peel, as you've no doubt guessed by now... floats.

Sandologist ON DUTY

## What's going on?

The peel of an orange acts like a little life jacket for the flesh which sinks without it.

Whether a substance will sink or float depends on its density. Density is how much a given volume of a substance weighs. For example, one litre of water weighs one kilogram. That's no coincidence by the way (see below). Sugar and salt are more dense than water so both will sink. Wood is less dense and so it floats.

Orange flesh is mainly water and contains natural sugars, which are more dense than water so the fleshy part of an orange sinks. The peel contains lots of air which is much less dense than water so it floats.

## How much does a kilogram weigh?

▼ The kilogram was originally based on the weight of one litre of pure water. That's still an excellent way to compare an unknown object's weight, but for very precise measurements, it has a few problems. The density of water changes with temperature, so one litre of cold water weighs a bit more than one litre of warm water. Changes in atmospheric pressure also affect water's density. So in 1889, an international prototype kilogram was made from an alloy of platinum and iridium. It is a cylinder with a height and diameter of 3.9 centimetres. The original is kept in France but copies were sent all over the world.

## 34 Mysterious Egg

Set this trick up secretly and everyone will be begging to know how you did it. You need pure salt with no anti-caking agents (check the ingredients on the package or ask your parents for help).

Half fill a glass with water and toss in an egg, gently though. Add a tablespoon of pure salt and stir until clear. The egg should float but add a bit more salt if it doesn't.

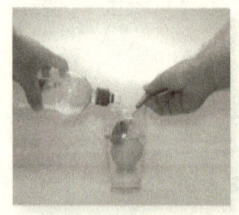

Gently pour fresh water over the back of a spoon until the glass is almost full.

Bizarro! The egg stays right in the middle of the glass. Push it to the bottom with a spoon and it bobs straight back to the middle again.

If you add a few drops of food colour and stir the top very gently, you'll end up with a layer of coloured water. Cool!

## What's going on?

serious experimentation underway

You already know that objects float if they're less dense than water. Eggs are more dense than fresh water so they sink. Adding salt increases the density of water. When the water becomes more dense than the egg, your egg floats.

Pouring fresh water over the spoon prevents it from mixing with the salty water below. The result is a layer of fresh water floating on top of the salty water. The egg floats on the salty water but sinks in the fresh water so it stays right in the middle. When salt dissolves, it becomes invisible so you can't tell it's there. Most table salt contains an anti-caking agent which doesn't dissolve in water.

## The Dead Sea

▼ Normal seawater contains about 35 grams of salt per litre on average. That's a lot but not quite enough to float you or an egg. The Dead Sea however, is landlocked between Jordan and Israel and much saltier than the open sea. Dead Sea surface water contains about 300 grams of salt per litre – almost ten times more than the oceans. This allows bathers to float on their back while reading the newspaper! You can enjoy the same experience at some meditation centres. Flotation tanks filled with super saline water allow you to float the same way. It's very relaxing but don't rub your eyes before drying them like I did… it stings like crazy which spoils the whole experience. Stupid floaty water!

# 35 Wasabi Diver

Wasabi is a deliciously explosive sauce that's yummy on fresh sushi. But a sachet of wasabi sauce also makes a mysterious and very obedient little pet-in-a-bottle.

You need an unopened sachet of wasabi sauce (available where you buy sushi) and a paperclip. This trick also works with some tomato sauce sachets, but not all.

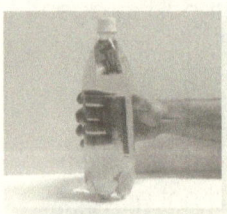
Attach a paperclip to the sachet and test whether it still floats, if not, just use the sachet by itself. Now place it inside a bottle filled to the brim with water.

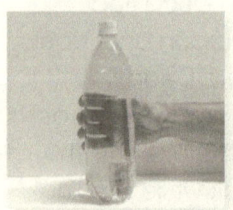
Screw the lid back onto the bottle and squeeze. The sachet sinks. Stop squeezing and it floats again.

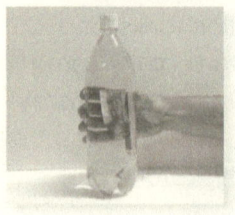
If you apply just the right amount of pressure, the sachet will hover anywhere you like. Well trained little sachet, eh? Sink sachet, sink! Good sachet.

SWIM BETWEEN
the experiments

### What's going on?

Wasabi sinks in water, but the sachet also contains a little air bubble inside which makes it float.

Water is an incompressible fluid which is a fancy way of saying you cannot squash water. Air on the other hand is easy to squash. So when you squeeze the bottle, you are actually compressing the little air bubble inside the sachet. Compressing the bubble reduces the buoyancy of your sachet so it sinks. When you stop squeezing, the bubble expands again and the satchet floats.

Adding a paperclip adds weight, which makes this trick a lot easier! Without the paperclip, you can still sink the sachet but you have to squeeze a lot harder.

### Wasabi's Chemical Punch

▼ The wasabi plant (Japanese horseradish) is a relative of mustard. Now plants can't run away from predators, so many have evolved a chemical 'punch-in-the-mouth' to defend against grazing animals. Wasabi contains a chemical called allyl isothiocyanate (AITC) which smacks animals right in the nervous system with a painful burning sensation. Garlic and chilli peppers have similar chemical weapons to ward off hungry mouths. Some people (like myself) love this sensation and the chemicals responsible even have medicinal qualities. So they're yummy and healthy!

# 36 Corrugated Paper

How much weight can a single sheet of paper support? It's a simple question with a not-so-simple answer which explains why people love their corrugated iron rooves.

Lay a sheet of paper across the gap between two bricks or stacks of books. Notice that it barely supports its own weight.

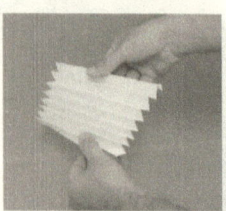

Now fold the paper back and forth as shown to make it corrugated. Make the folds about 1 cm apart.

Lay the paper across the gap and a coaster or a CD case on top. Start adding weights (eg: twenty-cent coins) into the plastic cup. How much do you think it can hold?

It took ten dollars and sixty cents in twenty-cent coins before my corrugated paper bridge collapsed. Amazing!

SCIENTISTS patrol these waters

## What's going on?

Folds or bends in a sheet of paper, metal or any other material can make it much stronger and more rigid. That's why corrugated iron is such a popular building material!

But the rigidity of a corrugated sheet is very directional. If you laid the paper with the corrugations running parallel to the bricks, it wouldn't hold much more than a standard flat sheet.

Corrugations also make flexible tubes like vacuum cleaner and creepy-crawler hoses more rigid to prevent them from being squashed.

### Accidental Cardboard Box

▼ Cardboard or 'corrugated fibreboard' is basically corrugated paper glued between two flat sheets. It was first invented in China and did not appear in the Western world until about 200 years later.

▼ Like so many other great inventions, the process used to mass-produce cardboard boxes was discovered by accident. In the late 1800s, printer and paper bag maker Robert Gair was preparing an order of seed bags. A ruler that was meant to crease the paper shifted and cut it instead. The accident taught Gair that cutting and creasing cardboard in one operation was not only possible but saved time. Boxes have been prefabricated using the process he invented ever since.

Adult HELP needed

# 37 Coin Popper

You'll cackle at the clinking of this classic science trick. It works with glass or plastic bottles so before you recycle them, tinkle them!

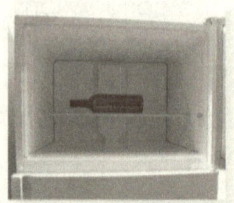
Put an empty bottle in the freezer for at least 30 minutes – don't rush, or this trick is a flopper, not a coin popper.

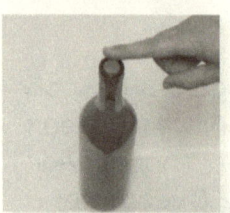
Moisten the mouth of the bottle with a little water using your finger.

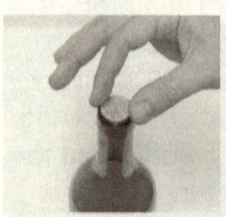
Carefully place a 10-cent coin on top of the bottle so it completely seals the hole.

Now wait, watch and listen carefully. After a second or two, you'll hear the coin go *clink*! A few seconds later, it will go clink again. And then again. If you look closely, you can even see it happening.

## What's going on?

Sandologist ON DUTY

The empty bottle wasn't really 'empty' when you put it in the freezer. It was full of air... *warm* air to be precise. When you took the bottle out, it was full of *cold air*.

Now here's a very important fact of nature. When you cool air down, it shrinks. When you heat air up, it expands.

As soon as you took the bottle out of the freezer, the cold air inside started to heat up. But the 10-cent coin blocking the hole prevents the air from expanding so the air builds up pressure instead. When the air pressure builds up enough, it can lift the coin for a brief moment, allowing a tiny bit of air escape. The clinking keeps going until the air in the bottle heats up to room temperature and stops expanding.

## Where waves come from

▼ Ocean waves are a result of the same process that makes your coin go 'clink'. No, I'm not kidding. Here's a brief rundown. Everyday, we spin in and out of sunlight. Sunlight heats everything up including air! When air heats up, it expands, so it rises. Cooler air flows in to replace the rising warm air. We call this airflow 'wind'. Wind blowing across water makes it wobble. If the wind blows hard enough, for long enough, these wobbles can grow into huge waves. Ocean waves keep travelling until they reach a shoreline where they eventually break. But it all started with the Sun heating air over a distant part of the ocean. Cool eh?

Adult HELP needed

# 38 Water Elevator

This candle-powered water elevator is sure to get a rise out of any audience. You need to blow fresh air into the bottle to replace the oxygen if want to do it again and again.

Float a tealight candle on a plate of water as shown. I used food colour so you can see the water clearly but it's not necessary.

Light the candle and let it burn for a few seconds, then cover it with a large jar or juice bottle.

Watch carefully – you'll see air bubbles escaping from the bottle.

When the candle goes out, the water rises up into the bottle taking the tea light candle with it. It's a candle-powered water elevator!

## What's going on?

The candle flame does two noticeable things when you cover it with a jar. First, it heats the air inside the bottle. When air heats up, it expands so some hot air escapes – that's what the bubbles were made of.

Another important thing is happening simultaneously. The candle flame slowly consumes the oxygen in the bottle. Soon there isn't enough left to sustain the flame, and it goes out.

When the flame goes out, it stops producing heat. Now the air inside the bottle starts cooling down. When air cools down, it contracts. This contraction lowers the air pressure inside the bottle. Outside the bottle, the pressure remains unchanged. Fluids always move from high to low pressure, so water gets pushed up into the bottle. It keeps rising until the pressure inside the bottle is equal to the pressure outside. Amazing!

serious experimentation underway

## The wrong explanation

▼ Here's a stunning example of how the most obvious explanations are not always correct. For many years, it was believed the water rose into the jar for a very different reason. It was thought that the missing oxygen consumed by the flame produced the lower pressure. Candle flames do consume oxygen, but in the process produce an equivalent volume of carbon dioxide gas and water vapour. It's the heat that does the trick demonstrating that even seemingly simple science can be deceptively complicated.

Adult HELP needed

# 39. April Fools' Banana

Here's a great trick for scientifically minded practical jokers. With a few big words and a lot of confidence, people will believe your bananas really are better than sliced bread!

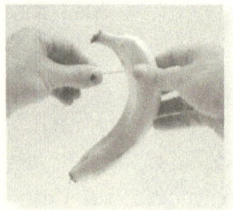

Do this in secret. Use a long darning needle or a straightened paperclip to cut into the flesh of an unpeeled banana.

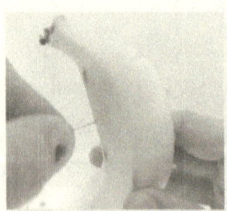

Carefully poke the needle into the banana as shown.

Wiggle the needle or paperclip sideways so it cuts the flesh. Repeat at equal intervals all the way down the banana.

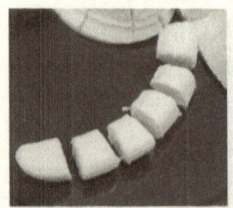

Now give the banana to a friend and watch their jaws drop as it neatly falls apart when they peel it.

SWIM BETWEEN
the experiments

## What's going on?

The secret to a really good April Fools' Day joke is a convincing story. Here's a suggestion for getting some extra mileage from your pre-sliced banana.

Tell your friend this is a brand new variety of genetically engineered banana developed in Coffs Harbour (home of the Big Banana). Now you need to bamboozle them with compelling facts including dates and scientific jargon! Tell them that scientists isolated the gene that causes citrus fruits (e.g. oranges, mandarins, lemons, etc) to grow in segments. Next they spliced this gene into banana DNA, resulting in a new genetically modified banana with flesh that grows in neat segments. It's a little more expensive of course, but think of the hours of chopping you'll save preparing fruit salads!

## Banana facts

- ▼ Bananas are real fruits, even though they don't have seeds. Scientists believe bananas lost their seeds about 10,000 years ago. Not lost as in 'oh dear, where'd I leave my seeds?' Lost as in a random mutation, which happens naturally in living things, all the time.
- ▼ Without seeds, new banana plants have to be grown from cuttings from their underground roots. Banana plants aren't trees. The trunk is made of tightly wrapped leaves, not wood.
- ▼ Bananas are a delicious source of potassium which is crucial for healthy brain function.

# 40 Slap on a Cap

This is the fastest and funniest way to put on a latex swimming cap! You can even do it to yourself, which should add a kooky touch to any serious swimming carnival.

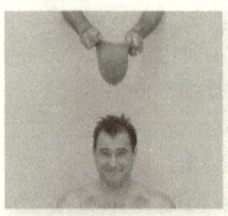

Fill a swim cap with water but don't over fill it. Hold the cap about 30 to 50 cm directly above your head.

Three, two, one...drop it. The water stays in the cap while it's falling.

I got my dad to drop this one while my mum giggled behind the camera – good shot dad!

When the cap hits your noggin, things get funny.

SCIENTISTS
patrol these waters

The water flows around the sides of your head, turning the cap inside out.

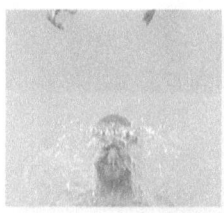

With the sides lowered, the water gushes out of the cap. This all happens in a split second and feels even funnier than it looks.

Ta-da! The latex cap snaps perfectly and snugly onto your scone.

You can't see a thing, of course, but who cares? It's hilarious!

### What's going on?
There's no need for a long-winded scientific explanation about gravity and the fluid properties of liquid water here. But it only works with the el-cheapo latex swimming caps. Forget those fancy-shmancy silicone caps because they don't work.

# Ocean Animal Facts

- ▼ The world's biggest animal is the blue whale, which can grow 30 metres long and weigh over 135 tonnes.
- ▼ The world's largest eyes are the size of a beach ball and belong to the mysterious giant squid.
- ▼ The world's most complicated eyes belong to the mantis shrimp.
- ▼ The world's largest invertebrate brain belongs to the octopus.
- ▼ The world's largest livers belong to sharks and they're rich in oil to increase buoyancy.
- ▼ Box jellyfish have twenty-four light sensitive eyes, some complete with a lens.
- ▼ Dolphin teeth are spaced just the right distance apart to detect underwater sounds in stereo.
- ▼ Turtles excrete excess salt through tear ducts in their eyes.
- ▼ Sea squirts eat their own brain once they've finished using it to find a suitable home.
- ▼ Fish scales have growth rings similar to those of a tree, which can be used to work out their age.
- ▼ Bony fish have a tiny balloon-like bladder, which they can fill with gas to adjust their buoyancy.
- ▼ Of the 400 species of shark, only four are considered a serious threat to humans: the bull shark, white shark, tiger shark, and oceanic white tip shark.

# Beach Facts

Sandologist ON DUTY

- ▼ Sandy beaches are home to more animal species than tropical rainforests.
- ▼ Of Australia's 11,011 beaches, South Australia's Coorong is the longest at 210 kilometres.
- ▼ Beach sand is mostly silicon dioxide (quartz), the most common substance in the Earth's crust.
- ▼ Giant beach worms, which live in sand near the low tide mark, can grow up to 2.5 metres long.
- ▼ Plastic bags take ten to twenty years to decompose in the ocean so any animal that dies from eating one will decompose before the bag!
- ▼ The weirdest flotsam and jetsam is ambergris (sperm whale vomit) – an amber-grey coloured substance used to make perfume and make-up.

# World's Weirdest Waves

The weirdest waves you can surf break in rivers, not at the beach. They come in twice a day and you can ride them for up to 10 kilometres upstream. They're called tidal bores and are made by the incoming tide.

They only occur where the tides get really big and can reach over three storeys high!

At certain times of the year, the Earth, Moon and Sun align to generate really big tides. In Australia, we call these King Tides and that's when tidal bores are most likely to form.

The shape of the river has to be just right for a tidal bore to form. You need a large, deep estuary leading into a narrow river, like a giant funnel. When the incoming tide rolls into the estuary, a huge body of water builds up momentum. When all this water gets squeezed into the narrow river, it builds up into a wall of water that breaks and even forms tubing sections along the banks.

Surfable tidal bores happen on the Seine River in France, the Severn in England and the Amazon in South America. The world's biggest tidal bores wreak havoc on the Qiantang River in China. We even get them here in the north of Australia on the Daly, Styx and Ord Rivers. Problem is we also have big crocodiles so surfing Aussie bores is a tad too risky for me.

# Safe experimenting

Sandologist ON DUTY

The science tricks in this book are easy, fun and safe. Please observe the following basic safety guidelines while you're experimenting:

### Water
▼ Do tricks with water outside or over the sink.
▼ Clean up spills because they can make smooth floors very slippery.

### Candles and matches
▼ Always ask an adult to help with candle experiments.
▼ Never leave a burning candle unattended.

### Food colours
▼ Food colour is non-toxic but can leave stains.
▼ Use only as much as suggested.
▼ Use an eye-dropper to add food colours.

### Plastic or glass?
▼ Where possible, use plastic jars and bottles instead of glass.
▼ Always ask an adult to help if you're using glass.

### Food
▼ Ask before you use food in your experiments.
▼ Compost scraps – don't bin them.

# Basic science trick equipment

Straws
Plastic cups
Toothpicks
Food colours
Detergent
Ground black pepper
Salt (*with no added anti-caking agent*)
Dinner plate
Balloons
Plastic bags (*water tight*)
Pencils
Aluminium cans (*empty*)
Water bombs
Ping-pong ball
Marble
Sticky tape
String
Large jar
Ice-cube tray
Tea bags

Matches – with adult assistance only!
Coins (5c, 10c, 20c, $2)
Hexagonal nut
Shoelace
Keys
Keyring
Coaster
Handkerchief
Polarised sunglasses
Paperclips
Wasabi in sachet (*from sushi shop*)
Paper
Tea light candle (*with adult assistance only!*)
Latex swim cap
Egg
Potato
Orange
Banana

# Safe experimenting

Sandologist
ON DUTY

The science tricks in this book are easy, fun and safe. Please observe the following basic safety guidelines while you're experimenting:

## Water
- ▼ Do tricks with water outside or over the sink.
- ▼ Clean up spills because they can make smooth floors very slippery.

## Candles and matches
- ▼ Always ask an adult to help with candle experiments.
- ▼ Never leave a burning candle unattended.

## Food colours
- ▼ Food colour is non-toxic but can leave stains.
- ▼ Use only as much as suggested.
- ▼ Use an eye-dropper to add food colours.

## Plastic or glass?
- ▼ Where possible, use plastic jars and bottles instead of glass.
- ▼ Always ask an adult to help if you're using glass.

## Food
- ▼ Ask before you use food in your experiments.
- ▼ Compost scraps – don't bin them.

# Basic science trick equipment

- Straws
- Plastic cups
- Toothpicks
- Food colours
- Detergent
- Ground black pepper
- Salt (*with no added anti-caking agent*)
- Dinner plate
- Balloons
- Plastic bags (*water tight*)
- Pencils
- Aluminium cans (*empty*)
- Water bombs
- Ping-pong ball
- Marble
- Sticky tape
- String
- Large jar
- Ice-cube tray
- Tea bags
- Matches – with adult assistance only!
- Coins (5c, 10c, 20c, $2)
- Hexagonal nut
- Shoelace
- Keys
- Keyring
- Coaster
- Handkerchief
- Polarised sunglasses
- Paperclips
- Wasabi in sachet (*from sushi shop*)
- Paper
- Tea light candle (*with adult assistance only!*)
- Latex swim cap
- Egg
- Potato
- Orange
- Banana

# Cool websites

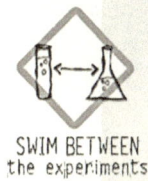

SWIM BETWEEN
the experiments

**ABC Science**
www.abc.net.au/science

**Roller Coaster**
www.abc.net.au/rollercoaster

**Questacon**
www.questacon.edu.au

**The Exploratorium**
www.exploratorium.edu

**NASA**
www.nasa.gov

**Bubble Rings – Dolphin Bubble Ring Sculpture Gallery**
www.earthtrust.org/delringgallery.html

**Surf checks, tides and webcams**
www.coastalwatch.com

**Bureau of Meteorology – national weather check**
www.bom.gov.au

**Finding words that rhyme**
www.rhymezone.com

**Converting measurements**
www.onlineconversion.com

**World population clock**
http://opr.princeton.edu/popclock/

Ruben Meerman is a surfer with a physics degree and a Grad Dip in Science Communication. He has taught primary science education at Griffith University, worked in the laser industry and performs hundreds of science shows schools throughout Australia every year. He appears on Triple J, ABC online, the ABC TV show *The ExperiMentals* and at a school near you!
www.abc.net.au/science/surfingscientist/
www.abc.net.au/experimentals

SCIENTISTS
patrol these waters

www.ingramcontent.com/pod-product-compliance
Lightning Source LLC
Chambersburg PA
CBHW022020290426
44109CB00015B/1250